Protecting Earth's Waters

by Cody Crane

Children's Press®
An imprint of Scholastic Inc.

Content Consultants
American Geosciences Institute

Library of Congress Cataloging-in-Publication Data Available

978-1-338-83716-2 (library binding) | 978-1-338-83718-6 (paperback)

10 9 8 7 6 5 4 3 2   23 24 25 26 27

Printed in China 62
First edition, 2023

Series produced by Spooky Cheetah Press
Book prototype and logo design by Book&Look
Page design by Kathleen Petelinsek, The Design Lab

Photos ©: 3 background and throughout: Freepik; 5 left: Paul Hobson/Minden Pictures; 5 right: Pollyana Ventura/Getty Images; 17: Wang Haibin/FeatureChina/AP Images; 19: Sergio Hanquet/Biosphoto/Alamy Images; 21: ufokim/Getty Images; 23: David Tonelson/Dreamstime; 24: SolStock/Getty Images; 28, 29 left, 29 right: Courtesy of Teach Beside Me; 30: AEDT/WENN.com/age fotostock.

All other photos © Shutterstock.

TABLE OF CONTENTS

Our Wet World.........................4

Chapter 1: Waters in Trouble............6

Chapter 2: Washed Away................10

Chapter 3: An Ocean of
                a Problem.....................16

Chapter 4: Cleaning Up.................22

Activity.................................28

Water Warriors...........................30

Glossary................................31

Index/About the Author..........32

Our Wet World

Take a look at Earth from space. All that blue you see is water. It covers most of our planet. This water is home to many creatures, such as corals and fish. Animals and plants on land rely on water, too.

In fact, all living things on Earth need water to survive. That includes people! It is important to protect this **natural resource**.

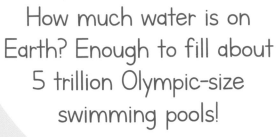

How much water is on Earth? Enough to fill about 5 trillion Olympic-size swimming pools!

5

Most of Earth's fresh water is found around the South Pole. It is trapped in the ice sheet that covers Antarctica. That is where many penguins live!

CHAPTER 1

Waters in Trouble

Nearly all the water on Earth is found in our oceans. It is salt water. We cannot use it for drinking or cooking. Only a small amount of Earth's water is fresh water. That is water that we can use. Most of Earth's fresh water is frozen. It is ice. Some is liquid. And it is in trouble.

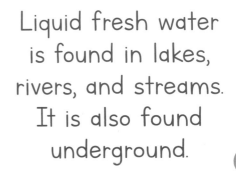

Liquid fresh water is found in lakes, rivers, and streams. It is also found underground.

One big problem is **pollution**. People create waste that can end up in **waterways** such as rivers and streams. Pollution harms all living things. It can also make water unsafe to drink.

Filtration can remove some pollution from water. Filters can be used to help clean the water used in people's homes.

Many people around the world do not have clean drinking water.

Organic fruits and vegetables are grown on farms that do not use harmful chemicals.

This plane is called a crop duster. It is spraying chemicals on a farm.

CHAPTER 2

Washed Away

How does pollution end up in water? Most of it starts on land. Many crops on farms are sprayed with chemicals. The chemicals kill insect pests and weeds. They help crops grow. Factories also use chemicals to make products like toys and cars. Harmful chemicals from farms and factories that are washed into waterways can become pollution.

Pollution comes from towns and cities, too. In these places, there is a lot of concrete covering the ground. Soil can soak up rainwater. Hard concrete cannot. Instead, as water flows over concrete, trash can mix with the water. Oil on streets can mix with the water, too. The trash and oil are kinds of pollution. The pollution runs into waterways.

People also use chemicals to help grass in lawns and parks grow. Rain washes the chemicals into waterways.

Concrete is the most used building material in cities.

Polluted water can be released into nearby waterways through large pipes.

Sewers are pipes that carry wastewater from toilets, sinks, and drains to a wastewater treatment plant. There the water is cleaned and released back into the environment. Sometimes, heavy rain makes a treatment plant overflow. Then the wastewater flows into nearby waterways. **Climate change** can cause more rain to fall, which can make treatment plants overflow more often.

CHAPTER 3

An Ocean of a Problem

Waterways can carry pollution from farms to the ocean. Some of this pollution is **fertilizer**. Tiny living things called algae live in oceans. Fertilizer can cause algae to grow fast. Oxygen in the water can be used up by the growing algae. That can cause other living things in the water to die.

Large growths of algae are called algal blooms.

Algae can cover large areas. These areas can be seen from space.

More than 500,000 garbage trucks worth of trash end up in the ocean every year.

Sea turtles sometimes mistake thin plastic for jellyfish. They try to eat the plastic.

Trash that people litter also makes its way to the ocean. Ocean animals can get tangled in the litter. They can get sick when they eat the trash.

Most of the trash is made of plastic. Plastic breaks down in nature very slowly. The plastic often crumbles into smaller and smaller pieces. The pieces can stay in the environment for hundreds of years!

Bits of plastic

Oil rig

Human activity can harm the ocean in other ways. Gasoline that we use to fuel our cars is made from oil. That oil is pumped from beneath the ocean by oil rigs. Sometimes oil rigs leak oil into the water. Ships carry oil around the globe. They can spill oil, too. Oil is hard to clean up. It can pollute the water and beaches for many years. It can kill ocean creatures.

Oil

This sea star is covered in oil from a spill in the ocean.

The city of Baltimore, Maryland, has a special barge. It scoops trash from the harbor.

CHAPTER 4

Cleaning Up

Now for the good news! Many governments are working to protect Earth's waters. Most countries have laws that limit water pollution from farms, factories, and cities. They have rules that ban littering. They work to lessen the effects of climate change.

Every year, experts meet during World Water Week. They discuss issues affecting the world's waters. Scientists study the health of Earth's waters to learn how to help.

People just like you are doing their part, too. They work to clean up beaches and waterways. They help raise awareness about water pollution.

Artists made sculptures from ocean trash. They want to show how much plastic pollution there is.

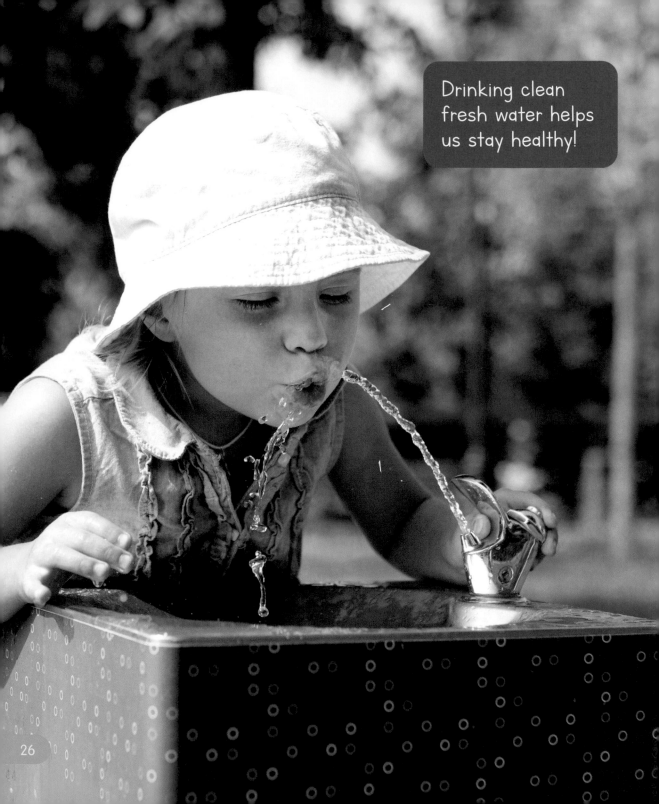

Drinking clean fresh water helps us stay healthy!

26

How can you help? Use fewer one-use items, such as plastic straws and bottled water. That will mean less trash that could end up as pollution. Find out more about your local water sources. Discover why they are important to your community. Learn how you can protect them. Remember, every living thing needs water. So do your part to protect Earth's waters!

FILTER OUT POLLUTION

Water can be filtered to help remove pollution. Try this activity to see how a filter can help clean up water.

WARNING! Do not drink any water in this experiment.

YOU WILL NEED

- An adult's help
- Plastic cup
- Scissors
- Two glass jars
- 10 coffee filters
- Sand
- Gravel
- Water
- Dirt, leaves, rocks, and sticks

STEPS

1 Have an adult poke a hole in the bottom of the plastic cup with the scissors. Set the cup inside the top of one jar.

2 Place five coffee filters inside the cup. Fill the top filter with sand.

3 Place another five coffee filters on top of the sand. Fill the top filter with gravel.

4 Pour tap water into the remaining jar. Mix in dirt, leaves, rocks, and sticks.

5 Slowly pour the water through the coffee filters. Look at the water that collects in the bottom of the jar. What happened?

6 Use the water in the jar to water your plants.

WATER WARRIORS

MEET MELATI AND ISABEL WIJSEN

Isabel (left) and Melati (right)

Melati and Isabel Wijsen live on the island of Bali. Bali is part of Indonesia, which is a country in Asia. When Melati was 12 and Isabel was 10, the sisters saw how plastic pollution was harming their island. Trash floated in the water and washed up on beaches. The girls started a movement called Bye Bye Plastic Bags. They got people to sign a request asking the government to ban plastic bags. The girls met with the island's governor. He agreed to ban plastic bags, as well as Styrofoam and plastic straws! Thanks to the Wijsen sisters, these items are no longer allowed on Bali.

GLOSSARY

climate change (KLYE-mit chaynj) long-lasting change to Earth's climate and weather patterns

fertilizer (FUR-tuh-li-zur) a substance added to soil to make plants grow better

natural resource (NACH-ur-uhl REE-sors) something that is found in nature and is valuable to humans, such as water or trees

pollution (puh-LOO-shuhn) substances that can be harmful to living things if added to water, land, or air

waterways (WAW-tur-wayz) channels for flowing water, such as rivers and streams

INDEX

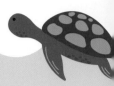

algae **16, 17**

animal **4, 19**

chemicals **10, 11, 13**

city **12, 13, 22, 23**

climate change **15, 23**

factory **11, 23**

farm **10, 11, 16, 23**

fertilizer **16**

filter **8, 28, 29**

fresh water **6, 7, 26**

ice **6, 7**

lake **7**

ocean **7, 16, 18–21, 25**

oil **12, 20, 21**

plastic **18, 19, 25, 27–30**

pollution **8, 11, 12, 16, 23–25, 27, 28, 30**

river **7, 8**

salt water **7**

sewer **15**

stream **7, 8**

trash **12, 18, 19, 22, 25, 27, 30**

treatment plant **15**

World Water Week **24**

ABOUT THE AUTHOR

Cody Crane is an award-winning nonfiction children's writer. From a young age, she was set on becoming a scientist. She later discovered that writing about science could be just as fun as the real thing. She lives in Houston, Texas, with her husband and son.